~A BINGO BOOK~

Fractions, Decimals and Percents Bingo Book

COMPLETE BINGO GAME IN A BOOK

Written By Rebecca Stark

ISBN 978-0-87386-427-5

Educational Books 'n' Bingo

Printed in the U.S.A.

FRACTIONS, DECIMALS & PERCENTS BINGO DIRECTIONS

INCLUDED:

List of Terms

Templates for Additional Terms and Clues

2 Clues per Term

30 Unique Bingo Cards

Markers

1. **Either cut apart the book or make copies of ALL the sheets. You might want to make an extra copy of the clue sheets to use for introduction and review. Keep the sheets in an envelope for easy reuse.**

2. Cut apart the call cards with terms and clues.

3. Pass out one bingo card per student. There are enough for a class of 30.

4. Pass out markers. You may cut apart the markers included in this book or use any other small items of your choice.

5. Decide whether or not you will require the entire card to be filled. Requiring the entire card to be filled provides a better review. However, if you have a short time to fill, you may prefer to have them do the just the border or some other format. Tell the class before you begin what is required.

6. There are 50 terms. Read the list before you begin. If there are any terms that have not been covered in class, you may want to read to the students the term and clues before you begin.

7. There is a blank space in the middle of each card. You can instruct the students to use it as a free space or you can write in answers to cover terms not included. Of course, in this case you would create your own clues. (Templates provided.)

8. Shuffle the cards and place them in a pile. Two or three clues are provided for each term. If you plan to play the game with the same group more than once, you might want to choose a different clue for each game. If not, you may choose to use more than one clue.

9. Be sure to keep the cards you have used for the present game in a separate pile. When a student calls, "Bingo," he or she will have to verify that the correct answers are on his or her card AND that the markers were placed in response to the proper questions. Pull out the cards that are on the student's card keeping them in the order they were used in the game. Read each clue as it was given and ask the student to identify the correct answer from his or her card.

10. If the student has the correct answers on the card AND has shown that they were marked in response to the *correct questions,* then that student is the winner and the game is over. If the student does not have the correct answers on the card OR he or she marked the answers in response to *the wrong questions,* then the game continues until there is a proper winner.

11. If you want to play again, reshuffle the cards and begin again.

Have fun!

TERMS

1/4	FRACTIONS
2/5	HIGHER-TERM FRACTIONS
9/16	HUNDREDS PLACE
3/5	HUNDREDTHS PLACE
3/4	IMPROPER FRACTION
.875	INTEGER
1 1/3	MINUEND
2 1/2	MIXED NUMBER
4 1/8	MULTIPLE
5 1/4	NUMERATOR
6	ONE
9.78	ONES PLACE
16	PERCENT
24.561	PLACE VALUE
32	PRODUCT
$34.88	PROPER FRACTION
48	RATIONAL NUMBER(S)
82	REDUCE
ADDENDS	RECIPROCAL
COMMON DENOMINATOR	SIMPLEST FORM
COMMON FACTOR	SUM
DECIMALS	TENS PLACE
DENOMINATOR	TENTHS PLACE
DIVISOR	THOUSANDTHS PLACE
EQUIVALENT FRACTIONS	WHOLE NUMBERS

Additional Terms

Choose as many additional terms as you would like and write them in the squares. Repeat each as desired.
Cut out the squares and randomly distribute them to the class.
Instruct the students to place their square on the center space of their card.

© Barbara M. Peller

Clues for Additional Terms

Write three clues for each of your additional terms.

_____ 1. 2. 3.	_____ 1. 2. 3.
_____ 1. 2. 3.	_____ 1. 2. 3.
_____ 1. 2. 3.	_____ 1. 2. 3.

50% **1/2**	**50%** **1/2**	**50%** **1/2**	**50%** **1/2**	**50%** **1/2**
50% **1/2**	**50%** **1/2**	**50%** **1/2**	**50%** **1/2**	**50%** **1/2**
50% **1/2**	**50%** **1/2**	**50%** **1/2**	**50%** **1/2**	**50%** **1/2**
50% **1/2**	**50%** **1/2**	**50%** **1/2**	**50%** **1/2**	**50%** **1/2**
50% **1/2**	**50%** **1/2**	**50%** **1/2**	**50%** **1/2**	**50%** **1/2**
50% **1/2**	**50%** **1/2**	**50%** **1/2**	**50%** **1/2**	**50%** **1/2**
50% **1/2**	**50%** **1/2**	**50%** **1/2**	**50%** **1/2**	**50%** **1/2**

1/4	**2/5**
1. .25 = ___	1. 2/3 x 3/5 = ___
2. 20/80 = ___	2. .40 = ___
3. 1/2 x 1/2 = ___	3. 40% = ___

9/16	**3/5**
1. 4/16 + 5/16 = ___	1. 4/5 x 3/4 = ___
2. 15/16 − 3/8 = ___	2. 4/5 ÷ 4/3 = ___
3. 3/4 x 3/4 = ___	3. .60 = ___

3/4	**.875**
1. 75% = ___	1. 14/16 = ___
2. 1/2 ÷ 2/3 = ___	2. 7/8 = ___
3. .75 = ___	3. 7/16 + 7/16 = ___

1 1/3	**2 1/2**
1. 7/9 + 5/9 = ___	1. 4 1/8 − 1 5/8 = ___
2. 12/9 = ___	2. 2.5 = ___
3. 2 7/9 − 1 4/9 = ___	3. 5/2 = ___

4 1/8	**5 1/4**
1. 2 7/8 + 1 1/4 = ___	1. 21/4 + = ___
2. 7 3/8 − 3 1/4 = ___	2. 5.25 = ___
3. 33/8 = ___	3. 16/4 + 1 1/4 = ___

Fractions, Decimals and Percents Bingo

6 1.　.4 ÷ 2/3 = ___ 2.　12/3 x 9 = ___ 3.　.6,000 ÷ 1.000 = ___	**9.78** 1.　.6.49 + 3.29 = ___ 2.　15.80 − 6.02 = ___ 3.　.3.26 x 3 = ___
16 1.　16 x 100 = ___ 2.　1.6 x 10 = ___ 3.　160 x .10 = ___	**24.561** 1. Twenty-four and five hundred sixty-one thousandths. 2.　24 561/1000 3. The long form of this number is (2 x 10) + (4 x 1) + (5 x 0.1) + (6 x .01) + (1 x .001)
32 1.　25% of 128 = ___ 2.　50% of 64 = ___ 3.　3,200 ÷ .100 = ___	**$34.88** 1.　$21.44 + $13.44 = ___ 2.　$50.34 − $15.46 = ___ 3.　$69.76 ÷ 2 = ___
48 1.　24 ÷ 1/2 = 2.　96 x 1/2 = 3.　.48 x 100	**82** 1.　25% of 328 2.　50% of 164 3.　10% of 820
ADDENDS 1. The numbers added in an addition problem are called ___. 2. In the mathematical statement 1/2 plus 1/4 = 3/4, 1/2 and 1/4 are these. 3. In the mathematical statement .5 plus .4 = .9, .5 and .4 are these.	**COMMON DENOMINATOR** 1. Two fractions with the same denominator are said to have a ___. 2. The lowest ___ of 1/2 and 1/4 is 4. 3. The lowest ___ of 1/2 and 1/3 is 6.

Fractions, Decimals and Percents Bingo

COMMON FACTOR
1. It is a number that can be divided into 2 or more numbers.
2. The greatest ___ is the largest number that can be divided into 2 or more numbers.
3. The greatest ___, or GCF, of 18 and 24 is 6.

DECIMALS
1. These fractional numbers use a mark similar to a period to express place value.
2. To convert these to percentages, multiply by 100.
3. They are numbers written to the base 10. Examples are .85 and 3.76.

DENOMINATOR
1. The numeral below the line in a fraction.
2. The numeral in a fraction that tells the number of equal parts into which the whole has been divided.
3. In the fraction 3/5. the numeral 5 is this.

DIVISOR
1. It is the number by which the dividend in a division problem is divided.
2. To divide fractions, you multiply by the reciprocal of this.
3. In the mathematical statement 86.24 ÷4 = 21.56, the ___ is 4.

EQUIVALENT FRACTIONS
1. These are fractions with an equal value.
2. 2/4 and 1/2 are ___ because they share the same value.
3. 6/8 and 3/4 are ___ because they share the same value.

FRACTIONS
1. A number that shows part of something.
2. Examples of proper ones are 3/4 and 2/3.
3. Examples of improper ones are 4/3 and 3/2.

HIGHER-TERM FRACTIONS
1. ___ have a numerator & denominator with a common factor greater than 1.
2. 5/10 and 12/18 are ___ because their denominators and numerators have a common factor greater than 1.
3. 6/8 is a ___ because the numerator and the denominator can be divided by 2.

HUNDREDS PLACE
1. The place three to the left of the decimal point.
2. In the number 548.621, the numeral 5 is in this place.
3. In the number 657.943, the numeral 6 is in this place.

HUNDREDTHS PLACE
1. The place just to the right of the tenths place and second to the right of the decimal point.
2. In the number 548.621, the numeral 2 is in this place.
3. In the number 657.943, the numeral 4 is in this place.

IMPROPER FRACTION
1. A fraction whose numerator is larger than its denominator.
2. A fraction whose denominator is smaller than its numerator.
3. 8/3 is this kind of fraction; so is 10/9.

Fractions, Decimals and Percents Bingo

© Barbara M. Peller

INTEGERS 1. The positive whole numbers, the negative whole numbers and zero. 2. These positive and negative numbers and zero can be shown on a number line. 3. The terms positive and negative can only apply to these.	**MINUEND** 1. It is the number from which another is subtracted. 2. In the mathematical statement .9 − .5 = .4, .9 is this. 2. In the mathematical statement .50 − .25 = .25, .50 is this.
MIXED NUMBER 1. It is part whole number and part fraction. 2. A positive ___ always has a value greater than 1. 3. 2 4/5 is one. So is 8 3/4.	**MULTIPLE** 1. A ___ of a number is the product of that number and any other whole number. 2. To find the lowest common denominator of fractions, find the lowest common ___ of the denominators. 3. The lowest common ___, or LCM, of the fractions 5/8 and 7/12 is 24.
NUMERATOR 1. The numeral above the line in a fraction. 2. The numeral in a fraction that tells the number of parts referred to. 3. In the fraction 3/5, the numeral 3 is one.	**ONE** 1. Any number times itself is that number. 2. 1/2 times this number is 1/2. 3. 101.5 times this number is 101.5.
ONES PLACE 1. The place just left of the decimal point. 2. In the number 548.621, the numeral 8 is in this place. 3. In the number 657.943, the numeral 7 is in this place.	**PERCENT** 1. This fractional number tells how many parts in a hundred there are in something. 2. To convert this to a decimal, divide by 100. 3. To convert a decimal to this, multiply by 100.
PLACE VALUE 1. Value given to a particular figure because of its place in the number. 2. In the number 34.987, the __ of 9 is "tenths." 3. In the number 34.987, the __ of 8 is "hundredths."	**PRODUCT** 1. It is the answer to a multiplication problem. 2. In the mathematical statement 5/8 x 2/5 = 1/4, 1/4 is this. 3. In the mathematical statement 2/5 x 10 1/5 = 4 1/5, 4 1/5 is this.

Fractions, Decimals and Percents Bingo

PROPER FRACTION
1. A fraction whose numerator is smaller than its denominator.
2. A fraction whose denominator is larger than its numerator.
3. 3/8 is this kind of fraction; so is 9/10.

RATIONAL NUMBER
1. It is a number that can be expressed as the ratio, or quotient, of 2 integers.
2. A number is called a ___ if we can write it as a fraction where the numerator and denominator are both integers.
3. Because 5 can be written 5/1, it is a ___ .

REDUCE
1. To change a higher-form fraction to a simpler form is to ___ the fraction.
2. To ___ a fraction to its simplest form, divide the numerator and the denominator by a common factor until their only common factor is 1.
3. We do this when we change 4/12 to 1/3.

RECIPROCAL
1. It is one of a pair of numbers whose product is one.
2. The ___ of 3/8 is 8/3.
3. The ___ of 4 1/2 is 2/9.

SIMPLEST FORM
1. The form of a fraction whose numerator and denominator have no common factor other than 1.
2. Fractions are in their ___ when they can be be reduced no further.
3. The fractions 2/4 and 3/9 are not in their ___ because they can be reduced further.

SUM
1. It is the result of adding two or more numbers together.
2. In the mathematical statement 1/2 plus 1/4 = 3/4, 3/4 is this.
3. In the mathematical statement .5 plus .4 = .9, .9 is this.

TENS PLACE
1. The place two to the left of the decimal point.
2. In the number 548.621, the numeral 4 is in this place.
3. In the number 657.943, the numeral 5 is in this place.

TENTHS PLACE
1. The place just to the right of the decimal point.
2. In the number 548.621, the numeral 6 is in this place.
3. In the number 657.943, the numeral 9 is in this place.

THOUSANDTHS PLACE
1. The place third to the right of the decimal point.
2. In the number 548.621, the numeral 1 is in this place.
3. In the number 657.943, the numeral 3 is in this place.

WHOLE NUMBERS
1. Positive numbers without a fraction or a decimal point.
2. 8, 12, 712 and 497 are ___. -8, 17.6 and 3/8 are not.
3. 9; 19; 1,416; and 34,687 are ___. -174, 16.7 and 0 are not.

Fractions, Decimals and Percents Bingo

Fractions, Decimals and Percents Bingo

Mixed Number	1/4	3/4	Denominator	Equivalent Fractions
Common Factor	2/5	Rational Number(s)	Addends	Hundredths Place
3/5	Thousandths Place		Product	Higher-Term Fractions
.875	24.561	Tenths Place	Multiple	Place Value
Percent	48	Decimals	Improper Fraction	Hundreds Place

Fractions, Decimals and Percents Bingo

.875	3/5	Minuend	Sum	9.78
Place Value	Addends	2 1/2	24.561	One
5 1/4	48		32	Tenths Place
Reciprocal	Simplest Form	Thousandths Place	Proper Fraction	Hundreds Place
Hundredths Place	Rational Number(s)	Decimals	Common Factor	Improper Fraction

Fractions, Decimals and Percents Bingo

.875	Tenths Place	Addends	Multiple	3/5
48	2/5	4 1/8	1/4	Fractions
24.561	Rational Number(s)		One	9/16
Thousandths Place	5 1/4	Percent	Reciprocal	Minuend
Improper Fraction	Common Factor	Decimals	Proper Fraction	9.78

Fractions, Decimals and Percents Bingo: Card No. 3

Fractions, Decimals and Percents Bingo

Thousandths Place	One	3/4	Common Factor	9.78
Integers	1 1/3	1/4	Sum	3/5
Product	Reciprocal		Equivalent Fractions	Denominator
Tenths Place	$34.88	Rational Number(s)	Decimals	2 1/2
Common denominator	Hundredths Place	Whole Numbers	Improper Fraction	Higher-Term Fractions

Fractions, Decimals and Percents Bingo

Hundredths Place	Equivalent Fractions	24.561	2 1/2	Common Factor
Integers	Tenths Place	4 1/8	32	2/5
3/4	Higher-Term Fractions		Numerator	82
Hundreds Place	9.78	Mixed Number	Proper Fraction	6
Addends	Decimals	3/5	Thousandths Place	Product

Fractions, Decimals and Percents Bingo

9/16	One	Minuend	9.78	Higher-Term Fractions
Multiple	24.561	6	1/4	3/5
Sum	Common Denominator		1 1/3	32
Decimals	Percent	Proper Fraction	Whole Numbers	3/4
Place Value	Tenths Place	Mixed Number	Product	$34.88

Fractions, Decimals and Percents Bingo

Mixed Number	One	82	Numerator	Addends
Place Value	9.78	48	2/5	Integers
Minuend	Denominator		32	1 1/3
Thousandths Place	Reciprocal	4 1/8	.875	5 1/4
Decimals	Common Factor	Proper Fraction	Whole Numbers	9/16

Fractions, Decimals and Percents Bingo

Product	One	16	Multiple	1 1/3
Integers	3/4	Sum	Higher-Term Fractions	2 1/2
$34.88	Reduce		9.78	Equivalent Fractions
Improper Fraction	Thousandths Place	.875	Common Denominator	Reciprocal
Rational Number(s)	Decimals	Whole Numbers	24.561	Place Value

Fractions, Decimals and Percents Bingo

32	Addends	48	$34.88	Common Factor
Common Denominator	9.78	Product	24.561	One
Fractions	Mixed Number		2/5	16
6	Hundreds Place	Percent	Numerator	82
Reciprocal	Proper Fraction	4 1/8	.875	Equivalent Fractions

Fractions, Decimals and Percents Bingo

.875	Multiple	1 1/3	Sum	$34.88
Higher-Term Fractions	2 1/2	1/4	2/5	9.78
Reduce	One		Denominator	5 1/4
Percent	Hundreds Place	6	Proper Fraction	Fractions
4 1/8	Place Value	Minuend	Hundredths Place	Product

Fractions, Decimals and Percents Bingo

9/16	One	24.561	6	Place Value
16	Fractions	Numerator	32	1/4
Integers	9.78		Minuend	48
4 1/8	3/5	Proper Fraction	Common Factor	.875
Common Denominator	Decimals	Mixed Number	Whole Numbers	Addends

Fractions, Decimals and Percents Bingo

Addends	Equivalent Fractions	Fractions	Multiple	32
48	Place Value	3/4	Whole Numbers	2/5
Mixed Number	82		Higher-Term Fractions	Sum
Decimals	Reciprocal	9.78	.875	Integers
One	16	Reduce	Common Denominator	2 1/2

Fractions, Decimals and Percents Bingo

6	Equivalent Fractions	9/16	Fractions	Higher-Term Fractions
3/4	16	9.78	32	5 1/4
Multiple	2 1/2		48	82
Product	Proper Fraction	1 1/3	Reduce	.875
Decimals	Hundreds Place	Whole Numbers	Mixed Number	Numerator

© Barbara M. Peller

Fractions, Decimals and Percents Bingo

Common Factor	9.78	24.561	32	Common Denominator
2 1/2	Mixed Number	Fractions	2/5	One
6	Denominator		Minuend	4 1/8
Hundreds Place	Proper Fraction	Reduce	1 1/3	9/16
Decimals	Sum	5 1/4	Place Value	Product

Fractions, Decimals and Percents Bingo

Numerator	32	24.561	Addends	Multiple
9/16	Minuend	1/4	3/4	Common Denominator
Higher-Term Fractions	Mixed Number		3/5	One
Decimals	Fractions	16	Proper Fraction	6
Place Value	Reciprocal	Whole Numbers	$34.88	48

Fractions, Decimals and Percents Bingo

1 1/3	Fractions	16	$34.88	Simplest Form
Sum	5 1/4	82	Integers	Denominator
6	Equivalent Fractions		Higher-Term Fractions	48
Thousandths Place	2 1/2	Decimals	Numerator	.875
Common Denominator	Tens Place	Whole Numbers	Reciprocal	One

Fractions, Decimals and Percents Bingo

4 1/8	Ones Place	Divisor	Fractions	Common Factor
Numerator	Common Denominator	Proper Fraction	Denominator	82
32	Product		Tens Place	16
Hundreds Place	Place Value	.875	24.561	5 1/4
Percent	6	Addends	Multiple	Equivalent Fractions

Fractions, Decimals and Percents Bingo: Card No. 17 © Barbara M. Peller

Fractions, Decimals and Percents Bingo

$34.88	Reduce	2 1/2	6	Sum
One	4 1/8	Percent	Higher-Term Fractions	Common Denominator
32	5 1/4		Divisor	3/4
Hundreds Place	1/4	Proper Fraction	.875	Minuend
Tens Place	Fractions	24.561	Ones Place	9/16

Fractions, Decimals and Percents Bingo

Higher-Term Fractions	9/16	Fractions	16	Reduce
Numerator	Multiple	One	Addends	Denominator
Ones Place	Common Factor		2/5	3/5
Minuend	Tens Place	Percent	Reciprocal	Divisor
3/4	Simplest Form	Place Value	Product	Whole Numbers

Fractions, Decimals and Percents Bingo: Card No. 19

Fractions, Decimals and Percents Bingo

Reduce	Ones Place	Multiple	Fractions	Whole Numbers
2 1/2	48	Integers	Percent	Sum
Equivalent Fractions	82		Thousandths Place	1/4
Hundredths Place	Rational Number(s)	Improper Fraction	Reciprocal	Tens Place
Tenths Place	Product	Simplest Form	.875	Divisor

Fractions, Decimals and Percents Bingo

Numerator	9/16	Integers	Fractions	Hundredths Place
Equivalent Fractions	Divisor	1 1/3	16	Mixed Number
5 1/4	Place Value		Ones Place	24.561
Percent	Addends	Tens Place	Hundreds Place	Product
Thousandths Place	Simplest Form	Whole Numbers	4 1/8	Reciprocal

Fractions, Decimals and Percents Bingo

$34.88	Minuend	Divisor	3/4	6
Sum	Multiple	3/5	16	2/5
2 1/2	Denominator		Mixed Number	82
Tens Place	Hundreds Place	Reciprocal	1/4	Common Factor
Simplest Form	4 1/8	Ones Place	5 1/4	Integers

Fractions, Decimals and Percents Bingo

1 1/3	Ones Place	Addends	3/4	Whole Numbers
9/16	Reduce	Place Value	Numerator	1/4
Minuend	6		Improper Fraction	Mixed Number
5 1/4	Simplest Form	Tens Place	4 1/8	Reciprocal
Hundredths Place	Rational Number(s)	Product	Percent	Divisor

Fractions, Decimals and Percents Bingo

1 1/3	Reduce	Common Factor	Ones Place	16
Divisor	Whole Numbers	Integers	Sum	Mixed Number
82	$34.88		6	5 1/4
Hundredths Place	Improper Fraction	Tens Place	4 1/8	Equivalent Fractions
Tenths Place	Thousandths Place	Simplest Form	Multiple	Rational Number(s)

Fractions, Decimals and Percents Bingo

Thousandths Place	Integers	Ones Place	24.561	Divisor
1/4	Hundreds Place	Numerator	1 1/3	2/5
Equivalent Fractions	16		Improper Fraction	Tens Place
3/5	Hundredths Place	Rational Number(s)	Simplest Form	Denominator
Whole Numbers	Common Factor	2 1/2	Common Denominator	Tenths Place

Fractions, Decimals and Percents Bingo

Divisor	Ones Place	Minuend	Sum	$34.88
Percent	Multiple	16	Reduce	1 1/3
Hundreds Place	Improper Fraction		Denominator	Thousandths Place
4 1/8	3/4	Hundredths Place	Simplest Form	Tens Place
82	Common Denominator	24.561	Rational Number(s)	Tenths Place

Fractions, Decimals and Percents Bingo

Minuend	2 1/2	Ones Place	Reduce	48
Hundredths Place	Improper Fraction	Numerator	Tens Place	2/5
Proper Fraction	Rational Number(s)		Simplest Form	Thousandths Place
$34.88	9/16	Integers	Tenths Place	1/4
Common Denominator	Denominator	Divisor	3/5	82

Fractions, Decimals and Percents Bingo

Higher-Term Fractions	Reduce	3/5	Ones Place	1 1/3
48	Divisor	Improper Fraction	Sum	Denominator
Rational Number(s)	5 1/4		82	Percent
.875	$34.88	Place Value	Simplest Form	Tens Place
3/4	32	Common Denominator	Tenths Place	Hundredths Place

Fractions, Decimals and Percents Bingo

Divisor	Reduce	$34.88	Numerator	32
Hundreds Place	Percent	Integers	82	3/5
Equivalent Fractions	Improper Fraction		2/5	Ones Place
48	Hundredths Place	9.78	Simplest Form	Tens Place
1 1/3	16	Tenths Place	9/16	Rational Number(s)

Fractions, Decimals and Percents Bingo: Card No. 29

Fractions, Decimals and Percents Bingo

Common Factor	Ones Place	Sum	32	Tens Place
1/4	Reduce	Minuend	Denominator	2/5
Hundreds Place	6		82	Integers
Tenths Place	9/16	3/4	Simplest Form	Improper Fraction
Hundredths Place	Addends	Rational Number(s)	Divisor	3/5

Fractions, Decimals and Percents Bingo: Card No. 30

www.ingramcontent.com/pod-product-compliance
Lightning Source LLC
Chambersburg PA
CBHW051419200326
41520CB00023B/7291